THE ENGLISH SUNRISE
Brian Rice and Tony Evans

Chatto & Windus. London

Published in 1986 by
Chatto & Windus Ltd
40 William IV Street
London WC2N 4DF

First published in Great Britain by Mathew Miller Dunbar 1972
Copyright © Brian Rice and Tony Evans 1972
Photographs copyright © Tony Evans 1972
Designed by David Hillman with Brian Rice and Tony Evans
Cover illustration: leaded window by Ray Bradley

ISBN 0 7011 3076 8

Printed by Optima Print and Packaging Ltd.

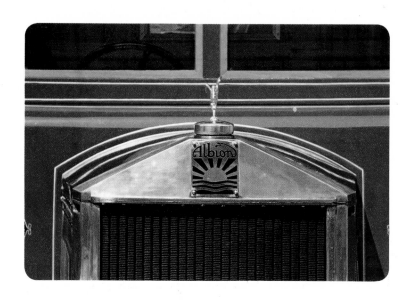

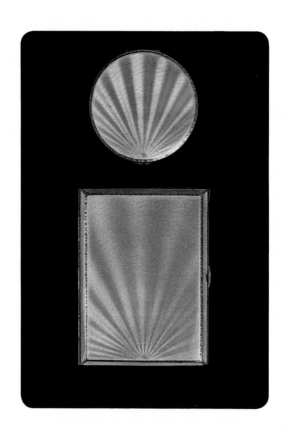

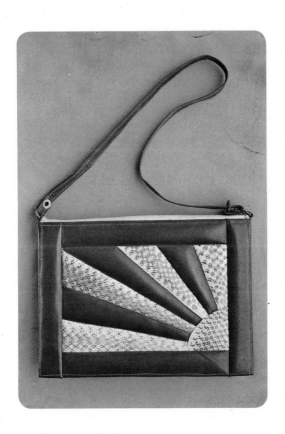

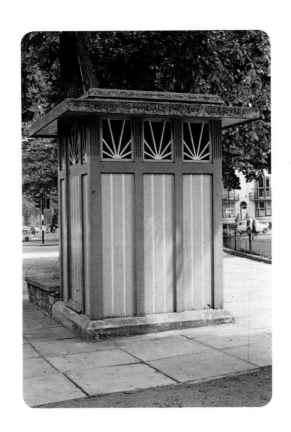

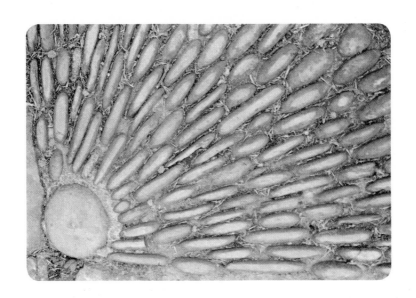

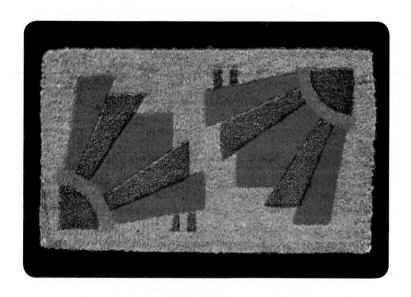

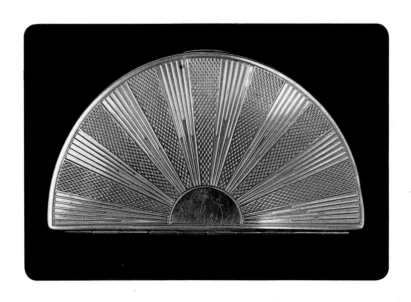

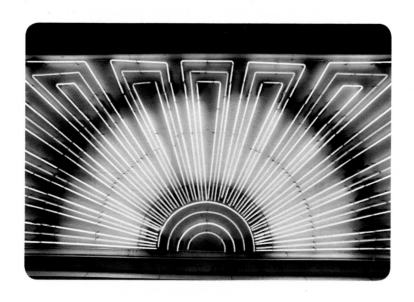

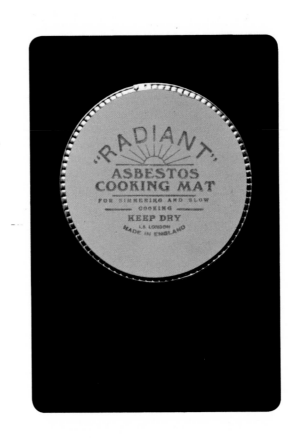

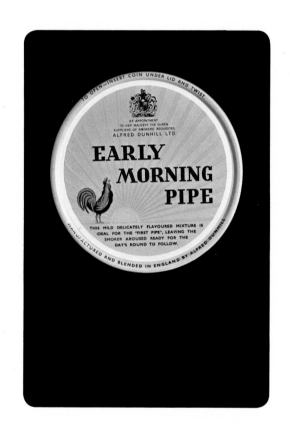

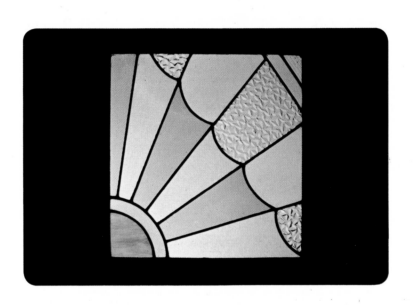

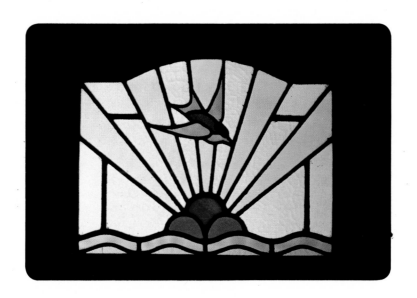

1 Shop door. *Crewkerne, Somerset.*
2 Gates. *Shaftesbury, Dorset.*
3 Radiator badge, Albion 30 cwt truck, 1926. *Collection G P Radcliffe.*
4 Badge, Sun bicycle.
5 Sign, Young Farmers' Club.
6 Post card, Western Counties Agricultural Association Ltd, 1927. *Collection Brian Rice.*
7 Cigarette case and powder compact. *Collection Harvey Daniels.*
8 Handbag. *Courtesy of Butler and Wilson, London.*
9 Door, former Wesleyan Chapel. *Lyme Regis, Dorset.*
10 Fanlight and surround, public house. *Birmingham, Warwickshire.*
11 Record label, Chappell and Co Ltd, London. *Collection Tony Evans.*
12 Gramophone needle tin. *Collection Tony Evans.*
13 House gable. *East Dereham, Norfolk.*
14 Kiosk. *Brighton, Sussex.*
15 Light fitting, restaurant car "Brighton Belle". *Courtesy British Railways.*
16 Porch lantern. *London.*
17 Relief panel, "Rising Sun" public house. *Wellingborough, Northants.*
18 Sign, Brickwoods public house. *Brading, Isle-of-Wight.*
19 Frontage, Regal Cinema. *Shanklin, Isle-of-Wight.*
20 Frontage, Royal Court Theatre. *London.*
21 Fuse box. *London.*
22 Radio. *Courtesy Chou, London.*
23 Frontage, butcher's shop. *Brighton, Sussex.*
24 Frontage, butcher's shop. *Brighton, Sussex.*
25 Gate and door. *Peasdown St. John, Somerset.*
26 Gate. *Near Ventnor, Isle-of-Wight.*
27 Chair. *Courtesy Christopher Baker, London.*
28 Chair. *Courtesy Lesley Frost, London.*
29 Fascia, decorating shop. *Birmingham, Warwickshire.*
30 Fascia, shop. *Newport, Isle-of-Wight.*
31 Sign, barber's shop. *Tewkesbury, Gloucestershire.*
32 Ventilator grill, shopfront. *Yeovil, Somerset.*
33 Window. *Nantwich, Cheshire.*
34 Window. *Waltham Cross, Hertfordshire.*
35 Cobblestones, Lloyd George's grave.* *Llanystumdwy, Caernarvonshire.*
36 Mosaic pavement, engineering showroom. *London.*
37 Emblem, lorry cabin.

*Welsh – sunrise.